Broadcasting and Politics

A SPEECH GIVEN TO

THE GUILD OF BRITISH NEWSPAPER EDITORS

AT COVENTRY BY

Ian Trethowan

DIRECTOR-GENERAL OF THE

BRITISH BROADCASTING CORPORATION

SATURDAY 22 OCTOBER 1977

Let me begin on a personal note and say how very pleased I am that this audience should be the one to hear – suffer, perhaps! – my first major address as the new Director-General of the BBC. It is, frankly, an awesome task I am taking on and it's good to begin among fellow-professionals. I like to think I remain a working journalist, certainly I am an editor – there's no ducking that – and assuredly it was provincial journalism that set both my professional standards and my sights.

Like the press, broadcasting is deeply engaged in the business of reporting politics, and of presenting to the electorate the democratic dialogue on which our society is built. Unlike the press, broadcasting is also itself an issue within politics. It is not only that we have regular committees of inquiry. The press too enjoys that interesting experience. Broadcasting, in addition, is to a considerable degree under the ultimate control of politicians.

Take the BBC. Our ultimate authority is a Board of Governors, appointed by the Crown, but obviously on the advice of the Government of the day. Our money comes from a licence fee, the level of which is fixed by the Government. Our broadcast frequencies – the means by which our radio and television signals are transmitted, the equivalent, as it were, of your printing presses – these too, are controlled by the Government of the day. You will understand, therefore, if we at times seem a little preoccupied with the question as to how we can safeguard our independence, how best we can make sure that our programme makers – our editors – enjoy a proper degree of editorial freedom, that freedom without which we cannot make our contribution to the processes of democracy.

Let me make clear that we are concerned with what might be rather than with what is. This is certainly true of the Board of Governors. Governments of both parties have so far been scrupulous in making sure, through the appointments, that the Board reflects a wide spectrum of opinion. In practice, in my experience, the Governors of the BBC have consciously seen themselves as an important part of the defence of our independence, and they have so acted. The fact remains that a Government with more sinister intentions could find in the power of appointment to the Board of Governors an uncomfortably convenient weapon.

The Government's control over broadcast frequencies, although very real, is probably inevitable. We have here a scarce and valuable national asset. Its allocation, therefore, must be a matter of public policy, and so for some form of public regulation. This has been found necessary in some form or other in every country in the world.

It is the Government's control of our finances of which, at the moment, we are most conscious, and most concerned. Historically, the licence system was devised in the 1920's as a means of underpinning the BBC's independence, of distancing our finances from direct Government supervision. The theory was that the licence would be fixed at a level which would last the BBC for a number of years, so that the Government of the day would not be able to use its control of our money to dictate our policy. In one area, of course, there was an exception – the External Services are financed by direct Government grant-in-aid. This, however, is not an argument for stretching the grant system to the whole of the BBC. On the contrary, the more we see how much the External Services are subject to the vagaries of short term political consideration, the more we are anxious to resist the grant system being extended to the rest of the BBC.

But this year we have been pushed a perilous step nearer such a system. The latest increase in the licence is designed to last for one year, and the Government knows it won't last for longer unless we cut back our services. We have, in other words, been given a cash limit, fixed at a level which is designed to force us to stay within the Government's guidelines on pay. I am concerned not only with our immediate financial problems – although they are obviously much in our minds – but equally with the wider principle that the independence of the public service broadcasting system in this country needs to have a source of money which is independent of day to day political considerations.

I am not for one moment claiming that the BBC should be exempt from the nation's economic problems. If there are sacrifices to be made, we should make our share. I would claim that, in fact, we have done so in pretty full measure. Over the past three years we have made repeated cuts – in real terms – in our programme budgets, and we have cut the amount of regular airtime. Not least, our staff have felt the full weight of the Government's incomes policy, to a degree which has left a substantial gap between their pay and that of their opposite numbers in commercial broadcasting. This is an intolerable situation, deeply unfair to the staff as individuals, and risking great damage to the quality of our output through the drain of experienced people.

I am not attempting to raise doubts about the objectives of the present Government. We recognise, that in their decisions about the licence fee, they have been guided by their preoccupations with the economy, and above all the rate of inflation. All those who know the present Home Secretary know him as a man of integrity, and liberal views – and the same is true of his senior officials. We are concerned, again, not with what is but with what might be.

The prospect of the BBC being financed by annual licence fee increases which would become indistinguishable from annual grants would obviously open the way to the possibility that some future Government, no doubt under extreme pressure, would seek to use its power over our purse to weaken our editorial independence.

You will, therefore, understand, I hope, if I have begun by approaching the subject of broadcasting and politics from the standpoint of independence. A broadcast journalist's approach to politics is basically no different from that of a print journalist, and I speak as one who has reported politics in both media. There are, of course, considerable technical differences, which in turn can mean that our problems are rather different from yours, but the principle is the same. We claim the right of any journalist to report to the electorate, national or local, what is being done by their representatives in their name – and, for that purpose, we need to be independent of any risk of editorial interference from those representatives.

I say report advisedly. We obviously have a duty to present the different sides of any argument on a political issue – you usually through your leader columns, we usually through our discussion programmes. But I believe strongly that our first responsibility to the public is to report the facts of any issue. I sometimes find a tendency in journalism today to short circuit this process – it is, I believe, called 'committed journalism', which I take to be a rather grand phrase for not bothering too much about the facts. Then we have 'investigative journalism'. I read in a newspaper recently that I'm supposed to be opposed to investigative journalism. On the contrary, I think investigation is what journalism is all about. What I don't understand is in what way investigative journalism is different from reporting. The answer, I suspect, is none – its just another grand phrase dreamed up by someone trying to con his editor into giving him a bigger office and more pay.

Broadcasters, then, no less than print journalists, see their first responsibility in politics as to report. This, however, has not always been the case, and I hope you will allow me to digress into a little history, to illustrate some of the problems that we have faced.

In the early days of broadcasting, the activities of its journalists were greatly circumscribed, and although we won a degree of freedom during the war, political journalism on the air remained heavily restricted right into the 1950s. Well into the '50s for instance, Parliament was still solemnly considering whether to keep the 'fourteen day' rule. That was a rule which prevented broadcasters from staging any discussion of an issue which was due to be discussed in Parliament over the succeeding 14 days. Then, as late as 1955, the reporting on the air of the general election was solely confined to the party political broadcasts. The sort of election reporting which filled the columns of the newspapers was prohibited on the air.

Some expansion of our journalism was achieved under the stewardship of Sir William Haley but two things above all changed the situation. The first was the coming of competition in broadcasting; the second was the arrival of Hugh Greene as Director-General of the BBC. By the time of the election of 1959, the broadcasters had won – or rather had seized – their independence as political journalists, and during the early 1960s there was an explosion of political journalism on the air. This, of course, created an obvious problem. We were trying, in a very few years, to develop the mature, sophisticated political journalism which newspapers had evolved over several generations. Inevitably we did not have sufficient people of experience – and inevitably, some things went wrong. Hugh Greene's successor, and my predecessor, Charles Curran, decided – I'm sure rightly – that we needed a period of consolidation, doubly so when we found ourselves involved in the problems of reporting Northern Ireland, the most difficult journalistic problem which has faced broadcasters, certainly since Suez.

But for all the difficulties, Hugh Greene's policy of lifting the BBC's journalism on to the same plane as that of print journalism was a landmark in the history of broadcasting in this country, and Charles Curran's period of consolidation has enabled us to grow the experience, and I hope the maturity, on which further to develop our coverage of political affairs.

I have stressed our concern for our editorial independence, because broadcasters certainly – and conceivably all journalists – should be careful not to take their freedom too much for granted. An article in *The Times* last year claimed that only one-fifth of the world's newspapers were independent, and I doubt whether the proportion would be any higher for broadcasters.

But we are sometimes asked, not unfairly, whether we accept that our freedom carries with it responsibilities. The answer is – yes, we recognise above all that our editorial independence must be seen within the context of a society based on representative democracy. A broadcast journalist's freedom is an integral part of that system: equally, that system is the only sure guarantee of his independence as a journalist. Freedom is not divisible. We cannot, in the final analysis, avoid recognising that we are part of a representative parliamentary democracy, operating under the rule of law. This doesn't mean we must wholly deny the air to the enemies of democracy, even when they are self-avowed. To do that would, of itself, be a restriction on democracy. But it does mean that broadcasters generally – and certainly the BBC as an institution – should consciously accept its position in the democratic system, and recognise the obligations which this position imposes. If I emphasise our recognition of our position in the democratic process, it's really for two reasons. First, we have in the past been accused of not recognising it – of arrogating to ourselves a position not only above party, but even above principle. We should be very clear that no broadcaster claiming editorial independence in a democracy can equally claim to be impartial between that democracy and any form of dictatorship.

Secondly, we have to recognise the special, and peculiar, power of broadcasting, particularly television. Every new form of communication is hailed as a revolution of infinite promise – and infinite danger. It was true of the Northcliffe revolution around the turn of the century, and of the spread of radio between the wars. Now we have television, and none of us who have worked in it are blind to its limitations as a means of conveying information: we recognise its ephemeral nature as compared with print, and the distraction of vision as compared with pure sound. But that having been said, we have to recognise the widely held belief that television is the biggest single influence on the minds of the people of this country at the present time.

This, in turn, must mean that television may well be a considerable influence on people's political attitudes. It is not the only influence, and there are many other factors in day to day life which will condition a person's attitude to how he is being governed, and how he casts his vote. Nonetheless, what he sees on 'the box' will be an important factor.

We can therefore scarcely be surprised if politicians are acutely sensitive as to how their affairs are covered on the air. One can see the position from their point of view. Here is this powerful medium of communication, the most important single influence on their constituents. Yet how they are presented on it is almost wholly out of their control, apart from the occasional official party broadcasts. So far as the main news and information programmes are concerned, the account which they give of political affairs is entirely in the hands of journalists who may represent a number of different programmes and organisations, but who are ultimately responsible to one of two public authorities. It is scarcely surprising that politicians – indeed, people in most walks of public life – tend to be hypersensitive towards broadcasters, in one hand a stopwatch, the other poised over the telephone. Charges of bias or triviality flash around, accusations of ignoring this great speech, or that percipient action.

What is the reality? It would be absurd to pretend that the broadcasting organisations are infallible, that mistakes don't happen. Yes, sometimes over a short period one party will get more appearances than the other, some worthwhile speech will fail to get the report it should, some minister or his shadow will fail to be interviewed. Even then, it is possible to be too simplistic. If one party is indulging in fierce in-fighting it will tend to chalk up more appearances on the air – but is that necessarily to the disadvantage of another party?

And the truth is that the broadcasters cannot move very far from the careful balance for which they strive without it being embarrassingly obvious, and without something being done to put it right. I've always thought that the preoccupation which political organisations show for the finer points of head-counting is a little exaggerated. Bias in sheer numbers is the easiest area for the broadcasters themselves to police.

What is more difficult to assess is something more subjective – the climate of opinion which television creates as a whole, its entertainment as well as its information programmes. People on the moderate side of politics have sometimes complained that television provides too gloomy a picture of society, that we place too much emphasis on what is wrong. People on the extreme side of politics on the other hand, complain of precisely the opposite: they say the picture we present is too bland, too comfortable, that, in the jargon phrase, we 'reinforce the consensus'.

In fact, we strive to preserve a balance. We believe it is right to show what is wrong – whether through reports in information programmes or through dramatisations. On the other hand, it is not our business to undermine what I think most of us recognise as a broad consensus as to values.

The problem of finding the balance emerges acutely when we come to violence on television. If we were to exclude all violence we would give, unhappily, a false picture of our society. On the other hand, if we over stress it we may risk encouraging it – may, we cannot be sure, the evidence is mixed. We are carefully studying the recent report of Dr Belson which does find at least the possibility of some relationship between television violence and juvenile crimes.

I mention violence because now – after Lewisham and Ladywood – we have to recognise it as a factor on the fringe of political life. It is, I believe, a small factor, but it is one which sets broadcasters a special difficulty. What, in effect, happens is that some small political group, with no hope of gathering more than a tiny fraction of the vote, sets out to create a violent confrontation which, if reported, would get it a prominence out of all proportion to the group's political importance. Yet if the confrontation takes place, and dozens of people are injured, including dozens of policemen, there is no question – we must report it, and by the way that television does report, by pictures. I won't pretend that this is not a difficult and worrying problem. On the one hand, there is the clear risk that the broadcasters are being manipulated by these small, unrepresentative fringe groups – that violent incidents are being staged to attract the publicity of the cameras. On the other hand, if the violence takes place, it must be reported. Somehow, we must tread a careful editorial path through these difficulties.

One thought which I find slightly reassuring is the argument that violence on the screen might not be damaging to the democratic process. On the contrary, it may – so the argument goes – so forcibly bring home to people the dangers of political violence, of mob politics, that it will actually reinforce people's preference for peaceful politics, the politics of discussion, the politics of the ballot box.

That, at least, is still the dominant form of politics in this country, and while we may welcome the fact, we in broadcasting have to recognise the difficulties which we face in trying to report, and present, the political dialogue. I have spoken of the power of television, and the sheer impact of the images projected into the corner of the drawing room is clear. But there is a danger that those images will be – not trivial, but over-simplified. We can address ourselves to some issue with the greatest dedication, and the greatest seriousness, but if we are to hold the attention of any given audience, what we present on the screen will tend to be more simplified than that audience will read in the equivalent newspaper. I say 'equivalent', because it would obviously be absurd to compare say a programme like *Nationwide* with a newspaper like *The Times*. But even comparing like with like, big audience programmes with big circulation papers, the problem remains.

I believe that broadcast journalists – and I must in justice include ITN as well as the BBC – are gradually beginning to work their way through this particular challenge. If you take the average broadcast news bulletin – even the half-hour ones – they cannot report as much as even the most attenuated popular newspaper. What they can do, at the expense of not getting very much involved in the more human interest stories, is to give a far greater proportion of their space to the major stories than will the equivalent newspaper. This means that, on the main stories with political implications, the news bulletins will certainly in relative terms give more than newspapers, and even in absolute terms. This is certainly true if you add to news bulletin coverage the attention given in the current affairs programmes – *Panorama, Tonight, This Week* and so on. Admittedly, these programmes get smaller audiences but equally some of the items aim at a more sophisticated audience than that of the more popular newspapers.

But there remains one glaring weakness in our coverage of the democratic debate – our reporting of Parliament. Television and radio, at their best, do not report at second hand: they transmit the event as it is happening. If they cannot do that – or choose not to – they can usually at least record the event, and transmit an edited version later. For broadcasting, an announcer reporting at second hand is second best.

This, however, is what we have hitherto been condemned to do with Parliament. The BBC has, of course, reported the activities of Parliament since the 1920s. In addition to the reports from Westminster in general news bulletins, we have *The Week in Westminster*, still going strong on radio, which started as far back as 1929; and I was myself privileged to start a companion television programme when BBC2 opened in 1964. But the most important development was the starting of *Today in Parliament* in 1945. For over 30 years BBC radio has broadcast a 15 minute summary both later the same night, and again the following morning. The BBC would always choose to do this, but it is not always realised that this is the one programme we have to broadcast by law. Clause 13 of the Licence and Agreement under which the BBC operates says that we 'shall broadcast an impartial account day by day prepared by professional reporters of the proceedings in both Houses of the United Kingdom Parliament'. That the account always has been impartial has, I'm glad to say, never been in dispute. We have not, to put it mildly, avoided charges of bias over some other programmes, but MPs of all parties have paid tribute to the fairness of *Today* and *Yesterday in Parliament*.

We are, nonetheless, reporting Parliament at second hand. Our first hand reporting of politics is curiously lopsided.

At election-time, we can transmit the event itself – the press conferences, the speeches, the walkabouts. At party conferences, we can transmit the event itself – the debates as they happen during the day, and then recorded highlights to larger audiences in the evening. But at the very heart of the democratic process, Parliament itself, all we have so far been able to transmit, ironically, is the opening ceremonial.

Now we are at least on the brink of one break through – on radio. Two years ago we carried out a four-week experiment of radio broadcasting from the Commons, and it was adjudged to have been a success. Occasionally we went over live, more often we presented later recorded extracts. Much of the success of the experiment stemmed from the extensive use made of Commons material by our local radio stations. Parliament itself having decided the experiment was a success, we had expected that permanent arrangements would be made within a matter of months. The process has in fact been somewhat leisurely, but now we believe we will be able to start permanent radio broadcasting from Parliament in the New Year. It will not be without its problems, particularly when we come to choose what debates to take live, and how much of them – there is a fear among back benchers, and among the smaller parties that there will be too much emphasis on the two front benches. It is a fair point, and one we shall watch. What is important, however, is that for the first time the public will be able to hear the democratic debate at its source – the floor of the House of Commons.

I think this is profoundly important, for the sake both of broadcasting, and of politics. What part of the Commons business is heard will still be at the judgment of the broadcasters, but the politicians will be heard in their proper and natural forum, not in a discussion mounted artificially in a studio.

From what I have said, you will not be surprised to know that I attach the greatest importance to getting the broadcasting of Parliament extended to television, and for very much the same reasons. I believe that it will end at least some of the tension which now exists between politicians and broadcasters.

I have seen it argued by senior politicians that televising the Commons would emphasize the bizarre or amusing or astonishing, and that MPs would seek to match these criteria. With respect, nothing like that has happened in the years we have televised the party conferences nor did it happen during the radio experiment in the Commons. Our experience of the party conferences is rather to the contrary, the parties have been over-concerned to present a favourable and responsible image.

I recognise, nonetheless, that MPs are very unlikely to agree to televising the House this side of the next election, and until they have had a chance to evaluate the radio experience. They have a number of legitimate anxieties, not least the fear that the presence of the cameras would physically alter the atmosphere. To the extent that these are technical questions, they can – and should – be dealt with. I have suggested that once the radio broadcasting of Parliament is under way, Parliament and the broadcasters should begin to extrapolate from this experience the lessons which might apply to television.

So much, then, for politics and broadcasting. One final word, if I may, on the wider business of journalism. Six months ago many of you will have heard my predecessor, Sir Charles Curran, pay tribute to the local press of this country and its editors. I warmly echo his words – and share his conviction that we in the BBC must maintain and improve our own services to local and regional communities if we are to fulfil our responsibilities as the national instrument of broadcasting.

We must, too – and we shall, I promise you – stand shoulder to shoulder with a free press in seeking not merely to maintain but constantly, by questioning, to improve the quality of our journalism. No doubt in the months and years ahead there will be many times when you will kick the BBC in general and me in particular. And, when we go astray – as no doubt we shall sometimes, though I hope not too often – so you should: we are there to serve the public, they pay for us, and we positively wish to be accountable to them. But I hope that at the same time you will recognise, as I do, that we also share one vital task: to provide the public with journalism which is accurate and authoritative, decent and determined, fair and stubbornly independent. On that we must surely be united.

Published by the British Broadcasting Corporation,
35 Marylebone High Street, London W1M 4AA,
and printed in England by Heffers Printers Ltd, Cambridge

ISBN 0 563 17484 6